...III

Mandar Aung

Some Important Paleozoic Fauna

Zaw Min Thein
Lae Lae Win
Thandar Aung

Some Important Paleozoic Fauna

in Zebingyi Formation, Kyaingtaung Formation, Nyaungbaw Formation and Thitsipin Limestone Formation of Myanmar

LAP LAMBERT Academic Publishing

Imprint

Any brand names and product names mentioned in this book are subject to trademark, brand or patent protection and are trademarks or registered trademarks of their respective holders. The use of brand names, product names, common names, trade names, product descriptions etc. even without a particular marking in this work is in no way to be construed to mean that such names may be regarded as unrestricted in respect of trademark and brand protection legislation and could thus be used by anyone.

Cover image: www.ingimage.com

Publisher:
LAP LAMBERT Academic Publishing
is a trademark of
International Book Market Service Ltd., member of OmniScriptum Publishing Group
17 Meldrum Street, Beau Bassin 71504, Mauritius

Printed at: see last page
ISBN: 978-620-0-46694-5

Table of Contents

Lists of Figures

Some Important Fauna of the Zebingyi Formation Exposed in the Kangyigon Area, Pyin Oo Lwin Township, Myanmar: Evidence for Middle Devonian (Eifelian)

Late Ordovician to Middle Devonian Fossils from the Nyanyintha Area, Pyin Oo Lwin Township, Mandalay Region

Microfacies Analysis of the Thitsipin Limestone Formation, Dattaw Area, Kyaukse District, Mandalay Region

Lists of Tables

Some Important Fauna of the Zebingyi Formation Exposed in the Kangyigon Area, Pyin Oo Lwin Township, Myanmar: Evidence for Middle Devonian (Eifelian)

Zaw Min Thein[1], Lae Lae Win[2], and Thandar Aung[3]

Abstract

The Devonian sediments are widely distributed in Kangyigon area, Pyin Oo Lwin Township. This area forming part of the western marginal zone of the Eastern Highland, is situated about 11.26 km south of Pyin Oo Lwin. The Zebingyi Formation is identified by its stratigraphic position: the lower black shale and limestone, and the upper buff to purple shale, siltstone and quartzose sandstone. The fauna from the lower part of the Zebingyi Formation are *Meristina* sp., *Monograptus* sp., *Nowakia* sp., *Styliolina* sp., *Dalmanites* sp., *Michelinoceras* sp., and the upper part contains *Nowakia* sp., *Styliolina* sp., *Dalmanites* sp., *Phacops taungtalonensis*, *Odontochile* sp., *Chonetes* sp., *Atrypa* sp., *Favosites* spp., *Heliophyllum* sp., *Zaphrentis* spp., *Eridophyllum* sp., *Cystiphyllum* sp., *Calceola* cf. *C. sandalina*, *Coenites* spp., etc. Based on the stratigraphic position and faunal evidences, the Zebingyi Formation can be assigned as the Early Devonian (Pragian) to Middle Devonian (Eifelian).

Keywords: *Calceola* cf. *C. sandalina*, Pragian, Eifelian, Zebingyi Formation, Kangyigon (Myanmar)

Introduction

The Paleozoic rocks are widely distributed in the northern Shan State. The area forming part of the western marginal zone of the Eastern Highland, is situated about 11.26 km south of Pyin Oo Lwin (Figure 1). The Zebingyi Beds described by La Touche (1913) is reinvestigated by IGCP (1980) and redefined as Zebingyi Formation which consists of various lithologic facies of limestone, siltstone and

1

shale. Myint Thein (1983) studied and mapped the units in the Kywetnapa-Letpangon area in Patheingyi and Maymyo Townships and his Zebingyi Formation has three distinct members, black limestone and black shale interbeds in the lower, marl and siltstone in the middle and quartzose sandstone in the upper.

Hang Khan Pau et al. (1993) subdivided into five lithostratigraphic units on the basic of limestone, shale and siltstone ratio. The Zebingyi Formation is reinvestigated and subdivided by Aye Ko Aung and Kyaw Min (2011), Aye Ko Aung (2012) into three members, viz., the lower Khinzo chaung Limestone Member, the middle Inni chaung Limestone Member and the upper Doganaing chaung Orthoquartzite Member.

The present study only follows up the widely accepted the stratigraphic name, "Zebingyi Formation" and subdivided into two members, viz., the lower black shale and limestone interbeds and the upper buff to purple shale, siltstone and quartzose sandstone (Zaw Min Thein, 1995).

Methodology
The first author has taken long field investigations at the Kangyigon area and its environs. The samples were hammered out from the outcrops of shale, siltstone and limestone. Fossils-bearing samples were systematically packed with tissues and tapes as necessary. Well preserved specimens were then labeled and photographed.

Geologic Setting
The Ordovician to Permian strata (Figure 2) is constituted of the Naungkangyi (Ordovician), Nyaungbaw Formation (Silurian), Zebingyi Formation (Early to Middle Devonian) and Maymyo Dolomite Formation (Late Devonian?). The area has distinct major structures; Nyannyintha anticline (overturn fold) and Kangyigon syncline.

2

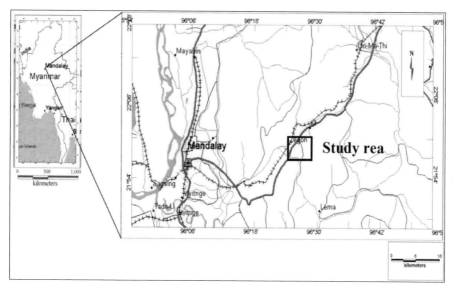

Figure (1) Location map of the study area

The Ordovician to Permian strata (Figure 2) is constituted of the Naungkangyi (Ordovician), Nyaungbaw Formation (Silurian), Zebingyi Formation (Early to Middle Devonian) and Maymyo Dolomite Formation (Late Devonian?). The area has distinct major structures, Nyannyintha anticline (overturn fold) and Kangyigon syncline.

Results

Distribution and Thickness

The Zebingyi Formation is limited in the western part of the area. It is mainly exposed in a narrow belt, trending north-south and dips about 20° to the east. The unit is well exposed in the vicinity of Dobin, Phaungdaw, Nyaungni and Thondaung (Figure 2).

The reference section of the formation, 102 m in the thickness, is located near the Kyinganaing chaung. The Zebingyi Formation has a gradational contact with the underlying Nyaungbaw Formation. It is unconformable overlain by the Maymyo Dolomite Formation of Late Devonian? age.

3

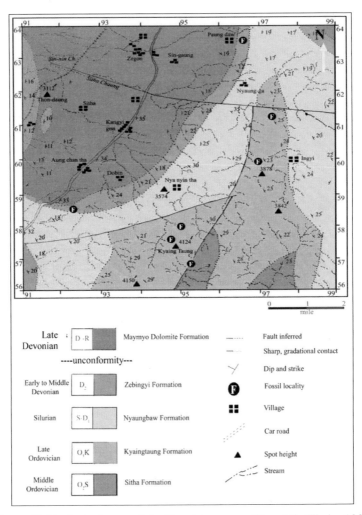

Figure (2) Geological map of the Kangyigon area (Zaw Min Thein, 1995)

Lithology

The Zebingyi Formation is identified by its stratigraphic position: the lower black shale and limestone, and the upper buff to purple shale, siltstone and quartzose sandstone. The contact between the Zebingyi Formtion and the underlying Nyaungbaw Formation is not observed in the Kangyigon area. The lower member

4

comprises 58 m of fine-grained, black limestone interbedded with black or carbonaceous shale showing well bedded nature (Figure 3-4).

The beds become thinner towards the top of the unit consisting tentaculites, graptolites, trilobites and brachiopods. The sequence of the lower member is overlain by the purple shale interbedded with buff color siltstone with trilobites and brachiopods.

It is succeeded by thin-bedded, reddish brown quartzose sandstone (Figure 5). It is 24 m thick and there are a lot of corals in these silicified units. The contact between Zebingyis and Maymyo Dolomite Formation is not observed in this area. It may be unconformable contact between them.

Figure (3) Thin-to medium-bedded, black shale and black limestone of the lower part of the Zebingyi Formation exposed in the Kyinganaing chaung showing thinning upward sequence

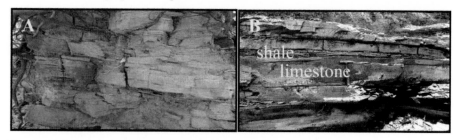

Figure (4) Black limestone interbedded with black or carbonaceous shale of the lower part of the Zebingyi

Figure (5) Thinly-bedded, buff colour siltstone or silty shale and reddish brown quartzose sandstone from the upper part of the Zebingyi Formation exposed in the Ingyi car road.

Fauna

The fossils tentaculites, trilobites (Dalmanites) are occurred throughout the whole sequence of the Zebingyi Formation. The important fossils are mollusks (Nowakia sp., Styliolina spp.), trilobites (Phacops taungtalonensis, Dalmanites sp., Odontochile sp.), brachiopods (Meristina sp., Atrypa sp., Chonetes sp.), graptolites (Monograptus spp.), cephalopods (Michelinoceras sp.), corals (Favosites spp., Coenites spp., Heliophyllum sp., Zaphrentis sp., Eridophyllum sp., Cystiphyllum sp., Calceola cf. C. *sandalina*) (Figures 6-9).

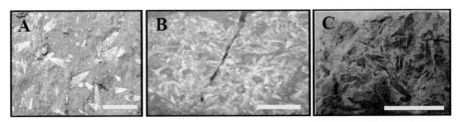

Figure (6) Fossils from the Zebingyi Formation: **A-C**. *Nowakia* sp. and *Styliolina* sp.; Scale bar is 1 cm.

6

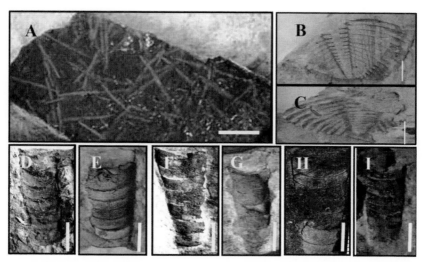

Figure (7) Fossils from the lower part of the Zebingyi Formation: **A**. *Monograptus* spp.; **B-C**. Pygidium of *Dalmanites* spp.; **D-I**. *Michelinoceras* spp. Scale bar is 1cm. (N 21° 57'10.577"; E 96° 25' 09.274")

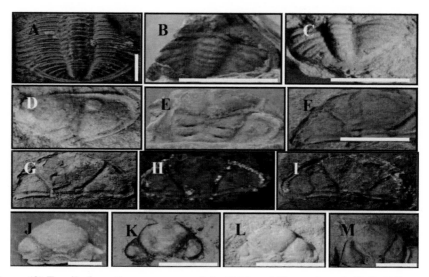

Figure (8) Fossils from the upper part of the Zebingyi Formation: A-C. Pygidiums of *Phacops taungtalonesis*; D&E. Cephalons of *Odontochile* sp., F-M. Cephalons of *Phacops taungtalonesis* (Scale bar is 1 cm)

7

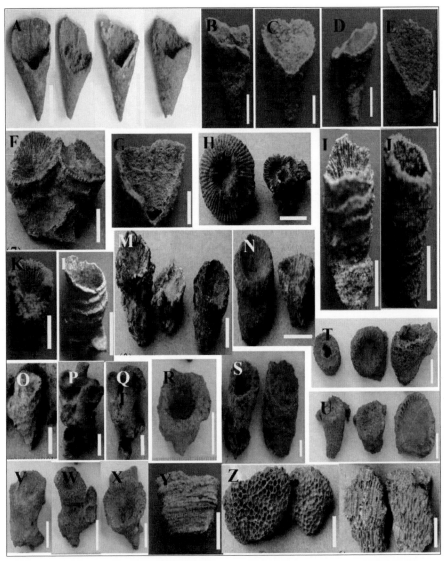

Figure (9) Fauna from the upper part of the Zebingyi Formation: A-E. Calceola cf. C. sandalina; F, I, L, S. Heliophyllum sp.; G, Y. Calceola sp.; H. Ptenophyllidae sp.; N, J, K, R, T-X. Zaphrentis sp. M, O. *Cystiphyllum* sp.; P-Q. ?*Pachyphyllum* sp.; Z. *Favosites* sp. Scale bar is 1 cm. (N 21° 57' 10.392"; E 96° 25' 12.323")

Discussion

Age

Abundant tentaculites (*Nowakia* sp. and *Styliolina* spp.) suggest an Early Devonian (Pragian) age. The graptolite fauna (*Monograptus* spp.) are suggested to the Early Devonian (Late Pragian-?Early Emsian) (Kyi Soe, 2000). The brachiopods (*Orthis* sp., *Meristina* sp., *Atrypa* sp., *Chonetes* sp.) and trilobites (*Phacops* spp., *Dalmanites.* sp.) assemblages are very similar to that of the Early Devonian in age. The trilobite species *Phacops taungtalonensis* (Thaw Tint and Hla Wai, 1970) discovered in the lower part of the Zebingyi Formation from Medaw area has been reported from Siegenian to Pragian. The condonts (*Eognathodus sulcatus*) a zonal form of *sulcatus* zone was recently discovered from the lower part of the Zebingyi Formation suggests a Pragian age.

Bo San (1985, personal communication) collected the Middle Devonian assemblages indicating *Favocities* sp., *Zaphrentis* sp. and *Coenites* sp. from the silicified rock layers. Lucky, the authors also found and collected the corals (*Favosites limitaris, Heliophyllum* sp., *Zaphrentis* sp., *Coenites* sp., *Calceola sandalina*) from the silicified rocks exposed near the Anisakan Landing Ground (N 21° 57' 10.392"; E 96° 25' 12.323") are suggested to the Middle Devonian (Eifelian) (Zaw Min Thein, 1995). The age of the Zebingyi Formation can be assigned as the Early Devonian (Pragian) to Middle Devonian (Eifelian) by the presence of conodonts, trilobites, brachiopods, tentaculites and corals.

Correlation

The tentaculites of the Zebingyi Formation are closely compared with those from the tentaculites-bearing Devonian units of Taunggyi-Taungchun range in Table 1 (Aye Ko Aung, 2010).

Table (1) Stratigraphic correlation of the Devonian units (Maung Thein, 2014)

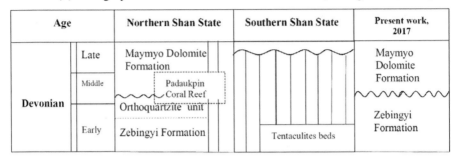

Age		Northern Shan State	Southern Shan State	Present work, 2017
Devonian	Late	Maymyo Dolomite Formation		Maymyo Dolomite Formation
	Middle	Padaukpin Coral Reef		
		Orthoquartzite unit		Zebingyi Formation
	Early	Zebingyi Formation	Tentaculites beds	

Conclusion

The Zebingyi Formation is predominantly composed of thin-to medium-bedded, black limestone and carbonaceous shale in the lower part. The upper part contains buff to purple, calcareous shale, siltstone and quartzose sandstone. Abundant tentaculites (*Nowakia* sp. and *Styliolina* spp.), graptolite (*Monograptus* spp.), brachiopods (*Orthis* sp., *Meristina* sp., *Atrypa* sp., *Chonetes* sp.) and trilobites (*Phacops* spp., *Dalmanites* sp., *Odontochile* sp.) assemblage are very similar to that of the Early Devonian in age. The corals (*Favosites limitaris*, *Heliophyllum* sp., *Zaphrentis* sp., *Coenites* sp., *Calceola sandalina*) from the silicified rocks exposed near the Anisakan Landing Ground (N 21° 57' 10.392"; E 96° 25' 12.323") are suggested to the Middle Devonian (Eifelian) (Zaw Min Thein, 1995). The age of Zebingyi Formation can be assigned as the Early Devonian (Pragian) to Middle Devonian (Eifelian) by the presence of conodonts, trilobites, brachiopods, tentaculites and corals.

Acknowledgements

The authors wish to express our sincere gratitude to Professor Hla Myint (retired) and Professor Bo San (retired) for their kind helps in various ways. Special thanks are also due to Dr Myint Thein (former Lecturer) for his invaluable advice and critical discussions.

References

Aye Ko Aung, 2010, A short note on the discovery of Early Devonian tentaculite-bearing unit from Taunggyi-Taungchun range, southern Shan State, Myanmar: Geological Society of Malaysia Abstracts with Programs, National Geoscience Conference 2010, p. 56.

Aye Ko Aung, 2012, The Paleozoic stratigrapgy of Shan Plateau, Mynamar-An updated version: Journal of the Myanmar Geosciences Society, Special volume, v. 5, no. 1 p. 1-73.

Hang Khan Pau, J., Ko Ko Gyi, and Aye Lwin. 1993, Microfacies analysis of the Zebingyi Formation: University of Mandalay Research Paper, [Mandalay], 33 p.

IGCP Burmese National Committee, 1980, Stratigraphic Committee Field Excursion in the Maymyo, Yadanatheingi, Hsipaw and Bawdwin areas: National Committee, p. 1-9.

Kyaw Min and Aye Ko Aung, 2010, New Early Devonian (Emsian) facies of Myanmar: Sub commission on Devonian Stratigraphy, Newsletter, 25, Münster University, p. 29-35.

Kyi Soe, 2000, Graptolites from the Zebingyi Formation, Zebingyi area, Pyin Oo Lwin Township: Paper read at Yangon University Paper Reading Ceremony (December, 2000), p. 1-7.

La Touche, T.H.D., 1913, Geology of the northern Shan State: Memoir of the Geological Survey [India], Calcutta v.39, 183p.

Maung Thein, 2014, Geological map of Myanmar, Explanatory Brochure, Myanmar Geosciences Society, Myanmar, 32 p.

Myint Thein, 1983, Geology of the Kywetnapa-Letpangon Area, Patheingyi and Maymyo Townships, [M.Sc. thesis]: University of Mandalay, 140 p.

Thaw Tint and Hla Wai, 1970. The Lower Devonian trilobite fauna from the east Medaw area, Maymyo District: Union of Burma, Journal of Science and Technology, v. 3, p. 283-306.

Zaw Min Thein, 1995, Geology and Stratigraphy of the Thapyegin-Nyaungni Area, Pyin Oo Lwin Township, [M.Sc. thesis]: University of Mandalay, 67 p.

Late Ordovician to Middle Devonian Fossils from the Nyanyintha Area, Pyin Oo Lwin Township, Mandalay Region

Zaw Min Thein[1*], Win Ei Zin Aung[2], Wai Yan Phyo[3], Thu Rein Soe[4]

Abstract

The study area is located about 11km south of Pyin Oo Lwin bounded by North latitude 21° 54' 00" and East Longitude 96° 21' 28" in one inch topographic map 93C/5. The present area is mainly composed of five lithostratigraphic units namely the Sitha Formation, the Kyaingtaung Formation, the Nyaungbaw Formation, the Zebingyi Formation and the Maymyo Dolomite Formation which are ranging in age from middle Ordovician to Devonian. The Kyaingtaung Formation is mainly composed of micaceous siltstone and marl. This siltstone and marl contain brachiopods, brozoans and crinoid steams. The Nyaungbaw Formation is mainly composed of medium to thick bedded, light grey to bluish grey, greenish grey, yellowish grey to reddish brown phacoidal silty lomestone with minor amount of calcareous siltstone, white shale and marl containing aboundant graptolites, brachipods, trilobites and crinoids stems. The Zebingyi Formation which consists of argillaceous limestone and black carbonaceous shale. In the study area, the lower part of Zebingyi Formation is mainly composed of thin to medium-bedded, black shale and black limestone. It is note that the calcareous shales of this unit are very fossiliferous with trilobites, graptolites, gastropods, cephalopods, brachiopods, *tentaculites* sp. and *styliolina* sp. The upper part of this formation is composed of reddish brown colored quartzose sandstone and thin bedded, purple or buff colored shale. This unit is very fossiliferous with corals, brachiopods, trilobites, *tentaculites* sp. *and styliolina* sp, palecypods and echinoids. In concluding, this finding is a new contribution to recent stratigraphy of Myanmar.

Keywords: Diplotripa westoni, Camerocrinus asiatus, Calceola cf. C. Sandalina, Ordovician, Silurian and Devonian

Introduction

The study area is located about 11km south of Pyin Oo Lwin bounded by North Latitude 21° 54' 00" and East Longitude 96° 21' 28" in one inch topographic map 93C/5 (Fig.1). This area lying in the western marginal zone of the Eastern Highland covers about 25 square miles of fairly rugged terrain. Since the study area is situated on the Mandalay-Pyin Oo Lwin motorway, it is easily accessible by car throughout the whole year.

Aims or Objectives

The main purposes of the present investigation are as follow;

1. To collect the index fossils and another important fossils
2. To designate the ages of rock units and
3. To interpret the stratigraphic position

Methodology

The first author has taken long field investigations at the Nyanyintha area and its environs. The samples were hammered out from the outcrops of shale, siltstone and limestone. Fossils-bearing samples were systematically packed with tissues and tapes as necessary. Well preserved specimens were then labeled and photographed.

Previous Work

Various geologic investigations of the northern Shan State were established by many geologists. Among them, La Touche (1913), Brown and Sondhi (1934), Chhibber (1934), Pascoe (1959), Brunnschweiller (1970) and Bender (1983) published the very remarkable stratigraphic succession.

Firstly, La Touche (1913) described the stratigraphic succession of Naungkhangyi area in Pyin Oo Lwin Township. Thaw Tint (1974) also stated the critical review of the Paleozoic stratigraphy of the northern Shan State and the new findings in the

Paleozoic paleontology. His emphatic area is Ngwetaung - Taunggyun - Kywenadauk area.

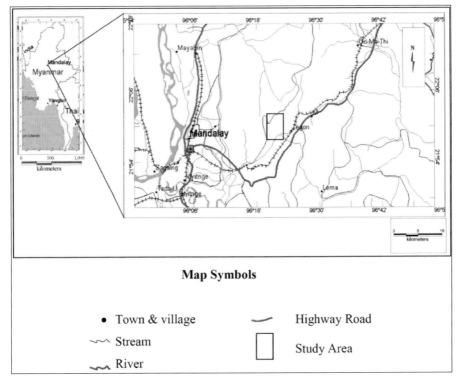

Figure (1) Location map of the study area

Previous Work

Various geologic investigations of the northern Shan State were established by many geologists. Among them, La Touche (1913), Brown and Sondhi (1934), Chhibber (1934), Pascoe (1959), Brunnschweiller (1970) and Bender (1983) published the very remarkable stratigraphic succession.

Firstly, La Touche (1913) described the stratigraphic succession of Naungkhangyi area in Pyin Oo Lwin Township. Thaw Tint (1974) also stated the critical review of the Paleozoic stratigraphy of the northern Shan State and the new findings in the

14

Paleozoic paleontology. His emphatic area is Ngwetaung - Taunggyun - Kywenadauk area.

Hla Myint (1978) mapped and described the geology of Paleozoic rocks of the adjacent area situated just west of the study area. Amos (1975) mentioned the stratigraphy of some of the Upper Paleozoic and Mesozoic carbonate rocks of the Eastern Highlands.

Win Win Kyi (1990) also mapped and described the geology of the Pathin - Kyadwinye area, lying in the eastern part of the area. The geology of the Kywenadauk-Okpho area located in the northern part of the study area was done by Ko Ko Gyi (1991). The geology and stratigraphy of Thapyegyin - Nyaungni area (including the present study area) was mapped by Zaw Min Thein (1995).

Regional Geologic Setting

The regional geologic setting of the area is shown in Figure (2). The study area lies in the western marginal zone of the Shan - Tanintharyi Block (Maung Thein, 1973). The present area is mainly composed of five lithostratigraphic units ranging in age from middle Ordovician to Late Devonian such as the Sitha Formation, the Kyaingtaung Formation, the Nyaungbaw Formation, the Zebingyi Formation and the Maymyo Dolomite Formation.

Stratigraphic Succession

In the present study area, the Paleozoic strata (Ordovician-Devonian) are exposed. Most of these outcrops are interbedded strata of carbonates and clastic rocks (mainly siltstone and shale). The stratigraphic nomenclature of the units in the Nyanyintha area follows up the previous nomenclature of all units which is in accordance with the stratigraphic practice. The stratigraphic succession the study area is shown in ascending order (Table 1).

15

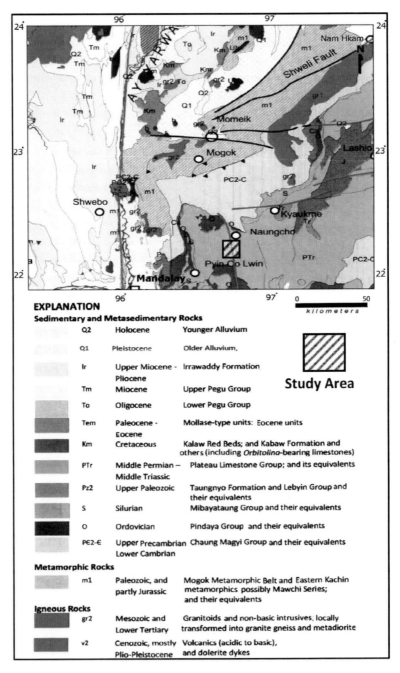

EXPLANATION

Sedimentary and Metasedimentary Rocks

Q2	Holocene	Younger Alluvium
Q1	Pleistocene	Older Alluvium,
Ir	Upper Miocene - Pliocene	Irrawaddy Formation
Tm	Miocene	Upper Pegu Group
To	Oligocene	Lower Pegu Group
Tem	Paleocene - Eocene	Mollase-type units: Eocene units
Km	Cretaceous	Kalaw Red Beds; and Kabaw Formation and others (including *Orbitolina*-bearing limestones)
PTr	Middle Permian - Middle Triassic	Plateau Limestone Group; and its equivalents
Pz2	Upper Paleozoic	Taungnyo Formation and Lebyin Group and their equivalents
S	Silurian	Mibayataung Group and their equivalents
O	Ordovician	Pindaya Group and their equivalents
PЄ2-Є	Upper Precambrian Lower Cambrian	Chaung Magyi Group and their equivalents

Metamorphic Rocks

m1	Paleozoic, and partly Jurassic	Mogok Metamorphic Belt and Eastern Kachin metamorphics possibly Mawchi Series; and their equivalents

Igneous Rocks

gr2	Mesozoic and Lower Tertiary	Granitoids and non-basic intrusives; locally transformed into granite gneiss and metadiorite
v2	Cenozoic, mostly Plio-Pleistocene	Volcanics (acidic to basic), and dolerite dykes

Study Area

0   50
kilometers

Figure (2) Regional geologic setting of the studied area and around it

Table (1) Stratigraphic succession the study area

	Formation	Age
5	Maymyo Dolomite Formation	Middle to Late Devonian
4	Zebingyi Formation	Early to Middle Devonian
3	Nyaungbaw Formation	Silurian
2	Kyaingtaung Formation	Late Ordovician
1	Sitha Formation	Middle Ordovician

Distribution

The Kyaingtaung Formation is well exposed at the Kyaingtaung (4174 ft). The Nyaungbaw Formation is widely developed in the middle and eastern part of the study area. This formation is well exposed in the vicinities of Nyanyintha village.

The Zebingyi Formation is limited to an extent in the middle part and- western part of the area. It is mainly exposed in the narrow belt. The good exposure of this unit can be seen along the Kyinganaing Chaung and Sitha Chaung. This unit can easily be seen in the vicinity of Dobin, Paungdaw, Naungni and Kyinganaing villages as shown in Figure (3).

Lithology

The Kyaingtaung Formation is very fossiliferous. The lower part of this formation is mainly composed of thin to medium-bedded yellowish-grey, calcareous siltstone, marl with lime parting.

Weathered surface shows black-stain. The upper part of this formation is composed dominantly of bluish-grey to greenish-blue silty limestone (Figure 4-a, b). It is also composed of buff to purple micaceous siltstone and marl. The lower member of Nyaungbaw Formation is mainly composed of medium to thick bedded, yellowish grey to reddish brown phacoidal, silty limestone with minor amount of calcareous siltstone, white shale and marl (Figure 5-a, b, Figure 6-a).

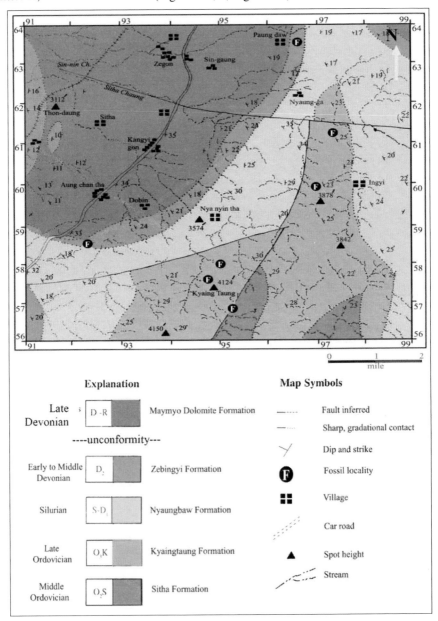

Figure (3) Geological map of the studied area (Zaw Min Thein, 1995)

(a)

(b)

Figure (4) (a) (b) Thin-bedded, buff to reddish-brown color siltstone of the Kyaingtaung Formation

(a)

(b)

Figure (5) (a) Bluish-grey to dark-grey colour of the argillaceous limestone showing the recumbent fold structure of the Nyaungbaw Formation, (b) Typical nodular bedding of the Nyaungbaw Formation

(a)

(b)

Figure (6) (a) Medium- to thick-bedded, grey-green and purplish colours of
argillaceous limestone with calcite veins of the Nyaungbaw Formation,
(b) Medium to thick-bedded, black limestone of the Zebingyi Formation
showing the opposite dip direction of bedding plane

In the study area, the lower part of Zebingyi Formation is mainly composed of the
thin to thick - bedded, black shale and black limestone which occur as narrow linear
belt (Figure 6-b). The shales are fissile and break into thin layers. It is noted that the

21

calcareous shales of this unit are very fossiliferous. The upper part of this formation is composed of the reddish brown colored quartzose sandstone and the thin bedded, purple or buff colored shale (Figures 7 and 8). In the present area, this formation is capped by the quartzose sandstones. This unit is also very fossiliferous.

Figure (7) Medium to thick - bedded, black shale and black limestone of the Zebingyi Formation showing the interbedded structure

Figure (8) Interbedded structure of thinly-bedded, buff to purple colour siltstone and reddish brown colour quartzose sandstones from the upper part of Zebingyi Formation.

22

Fauna

The important fossils of Kyaingtaung Formation are Brachiopods (*Saucrorthis* spp., *Nicolella* spp., *Stroptromena* sp.? and *Plaesiomy* sp.), Bryozoans *(Diplotrypa westoni)*, Ech-inoids (*Carocrinus* spp.?, *Xenocrinus* sp., Crinoid stem, *Cyclocaudex typicus* and Crinoid ossiles). The fossils collected from the Nyaungbaw Formation are Echinoids (*Camerocrinus Asisticus*), Cephalopods (*Michelinoceras* spp.), Graptolites (*Pseudoretiolites periatus; Petalolithus praecursor;, Campograptus rostratus,,. Campograptus lobiferous;, Neodiplograptus magnus,. Pristiograptus concinnus, Rastrites geinitzii, Petalolithus ovatoelongatus, Metaclimacograptus* sp.,*Demirastrites triangulates, Campograptus rostratus, Climacograptus* sp,, *Torquigraptus decipiens and Torquigraptus decipiens.*).

The most important fossils of Zebingyi Formation are Molluscas (*Tentaculites* sp., and *Styliolina* sp.), Cephalopods (*Michelinoceras* spp.), Brachiopods (*Orthis* spp , *Meristina* sp and *Chonetes* sp.), Graptolites (*Monograptus* spp, *Climacograptus* spp, and *Diplograptus* spp.), Corals (*Zaphrentis* spp, *Heliophyllum* spp,*Cystiphyllum* spp, *Calceola* cf. *C Sandalina, Tryplasma Lonsdale,Coenites* spp, *Favosites* spp and *Eridophyllum* sp.), Trilobites (*Phacops Taungtalonensis and Odontochile* sp.*) and* Cirripedia (*Palaeocreusia).*

Some important fossils from the Kyaingtaung Formation are shown in Figure (9) while some important fossils from the Nyaungbaw Formation are shown in Figures (10) and (11), and fossils from Zebingyi Formation are in Figures (12), (13), (14) and (15).

Figure (9) Fossils from the Kyaingtaung Formation: A, *Saucrorthis* sp., B-D, *Nicolella* spp., E, *Plaesiomy*? sp., F, *Diplotripa westoni.*, G&H, . *Carocrinus?* spp., I. *Xenocrinus* sp., J, Crinoid stem., K, Crinoid ossiles, L, Bryozoans.

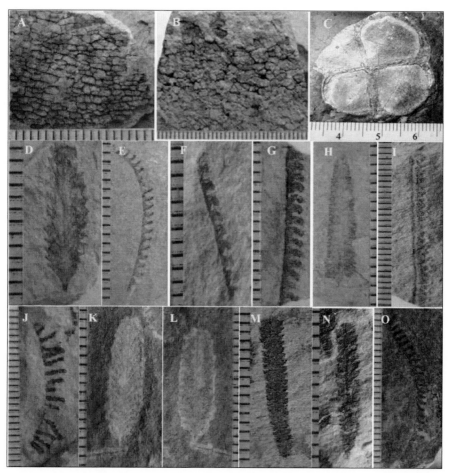

Figure (10) Graptolites from the Nyaungbaw Formation: A, B; Pseudoretiolites periatus Nyaungbaw, C; Camerocrinus asiaticus, D; Petalolithus praecursor, E; Campograptus rostratus, F,G; Campograptus lobiferous, H; Neodiplograptus magnus, I; Pristiograptus concinnus, J; Rastrites geinitzii, K, L; Petalolithus ovatoelongatus, M, N; Metaclimacograptus sp., O; Demirastrites triangulates

25

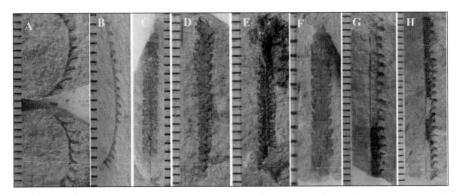

Figure (11) Some more Graptolites from the Nyaungbaw Formation: A, B;
Campograptus rostratus, C; *Climacograptus* sp., D, E, F;
Metaclimacograptus sp., G, H; *Torquigraptus decipiens*

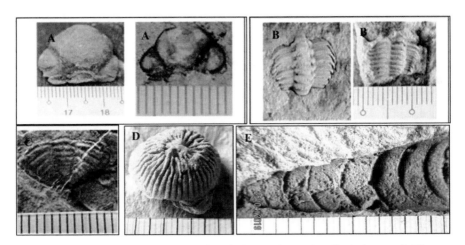

Figure (12) Fossils from the Zebingyi Formation: A; Cephalons of Phacops
Taungtalonensis, B; Thoraxic of Phacops Taungtalonensis, C;
Pygidiums of Phacops Taungtalonensis, D; Palaeocreusia, E;
Michelinoceras sp.

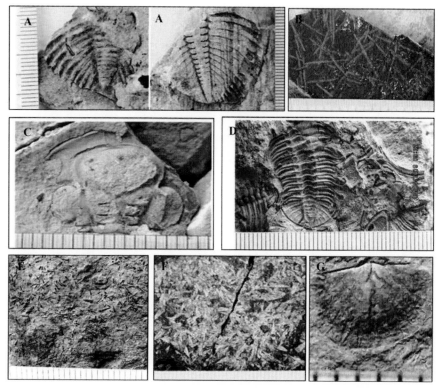

Figure (13) More fossils from the Zebingyi Formation: A; Pygidiums of *Odontochile* sp., B; *Monograptus* spp., C; Cephalons of *Odontochile* sp., D; *Phacops Taungtalonensis,* E; *Tentaculites* sp. F; *Styliolina* sp., G; Chonetes sp.

Figure (14) *Calceola* cf. *C. Sandalina* Corals from the Zebingyi Formation

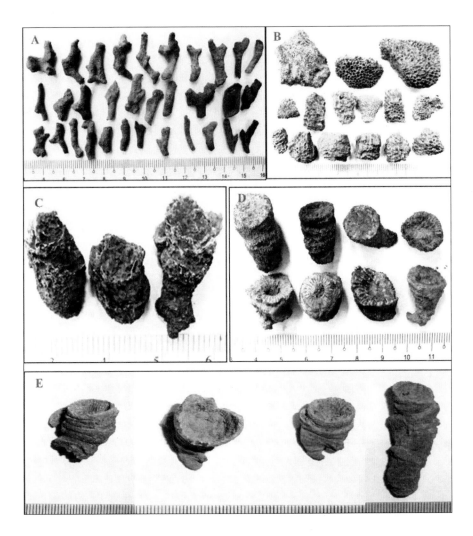

Figure (15) Corals from the Zebingyi Formation; A; *Coenites* spp., B; *Favosites* spp,
C; *Cystiphyllum* spp., D; *Zaphrenthis* spp., E; *Helliophyllum* spp.

Discussion

Fauna, Age and Correlation

The collected fossils of Kyaingtaung Formation indicate primarily Late Ordoviciam
age. The Kyaingtaung Formation of the study area can be correlated with the Upper

Naungkangyi Series of La Touche (1913), the Kyaingtaung Formation of I.G.C.P (1980) and the Nan-on Formation of Pindaya Group (Myint Lwin Thein, 1973).

These faunal assemblages of Nyaungbaw Formation indicate that the age of Nyaungbaw Formation is most probably Silurian in age. This unit of the study area can be correlated with the Panghsapye beds of La Touch (1913). He is also correlated with the Linw e Formation of Myint Lwin Thein (1973) in southern Shan State.

Thaw Tint and Hla Win (1970) discovered the trilobite (*Phacops taungtalononsis, Odontochile* sp.) from the lower part of Zebingyi Formation. Aye Ko Aung (2012) discovered the conodonts (*Eognsthodus Sulcatus*) from the lower part of Zebingyi Formation.

Luckily, it is also found that the collected corals (*Favosites Limitavis, Heliophyllum* sp., *Zaphrentis* sp. *Coenites* sp. *Calceola Sandalina*) from the upper part of Zebingyi Formation are suggested to the Middle Devonian (Eifelian) (Zaw Min Thein, 1995). Therefore the age of the Zebingyi Formation can be assigned as the Early Devonian (Pragian) to Middle Devonian (Eifelian) by the presence of conodonts, trilobites, tentaculites corals. This formation can be correlated with the tentaculites – bearing Devonian units of Taunggyi-Taungchun range (Aye Ko Aung, 2010).

Conclusion

The Kyaingtaung Formation is very fossiliferous. The lower part of this formation is mainly composed of the thin to medium-bedded yellowish-grey, calcareous siltstone, marl with lime parting. The important fossils of Kyaingtaung Formation are Brachiopods, Bryozoans, Echinoids. The above fossils indicate primarily Late Ordoviciam age.

The Nyaungbaw Formation is mainly composed of the thin to medium bedded, yellowish grey to reddish brown phacoidal, silty limestone and bluish grey to grey silty limestones. The fossils collected from the Nyaungbaw Formation are

Cephalopods, Echinoids and Graptolites. These faunal assemblages indicate that the age of Nyaungbaw Formation is most probably Silurian in age.

The Zebingyi Formation is mainly composed of the thin to thick - bedded, black shale and black limestone and reddish brown colored quartzose sandstone and thin bedded, purple or buff colored shale. The most important fossils of Zebingyi Formation are Molluscas, Cephal- opods, Brachiopods, Graptolites, Gastropod, Corals, Trilobites and Cirripedia. The age of the Zebingyi Formation can be assigned as the Early Devonian (Pragian) to Middle Devonian (Eifelian).

Acknowledgement

We would like to express our gratitude to Dr. Ba Han, Rector of Meiktila University, Dr. Kay Thi Thin, Pro-Rector of Meiktila University for their encouragements.

References

Amos, B.J. 1975. Stratigraphy of some Upper Paleozoic and Mesozoic carbonate rocks of the Eastern Highlands, Newsletter of stratigraphy, 4 (1); 49-70.

Aye Ko Aung, 2010. Aye Ko Aung, 2010, A short note on the discovery of Early Devonian tentaculite-bearing unit from Taunggyi-Taungchun range, southern Shan State, Myanmar: Geological Society of Malaysia Abstracts with Programs, National Geoscience Conference 2010, p. 56.

Aye Ko Aung, 2012, The Paleozoic stratigrapgy of Shan Plateau, Mynamar-An updated version: Journal of the Myanmar Geosciences Society, Special volume, v. 5, no. 1 p. 1-73.

Bender, F., 1983. *Geology of Burma*; Gebruiider Borntraeger, Berlin-Stuttgart.

Brown J. and V.P. Sondhi, 1934. The geology of the country between Kalaw and Taunggyi. S.S.S. *Rec. Geol. Surv. India*, V.67, p-166-248.

Brunnschweiler, R.O, 1970. Contributions to the post-Silurain geology of Burma (Northern Shan States and Karen State). *Jour.geol.soc.* Australia, 17:59-97, 10 Fig; Sydrey.

Chibber, H.L., 1934. *Geology of Burma*. Macmillan, London, 538p.

Compton, R.R., 1985. *Geology in the Field*, John Willy and Sons, Inc., U.S.A. 398p.

Hla Myint, 1984. Geology of the Ngwetaung-Taunggyun Area, Patheringyi and Pyin Oo Lwin Township; unpublished M.Sc Thesis, Geol. Dept., University of Mandalay. 106pp.

I.G.C.P., 1980. Stratigraphic Committee Field Excursion in the Maymyo, Yadanatheringi, Hsipaw and Bawdwin Areas, National Committee, Unpublished report. Pp. 1-9.

Ko Ko Gyi, 1991. Geology and minerals resources of the Kywenadauk-Okpho area, Pyin Oo Lwin Township; unpublished M.Sc Thesis, Geol. Dept., University of Mandalay. 106pp.

La Touche, T.H.D, 1913. Geology of the Northern Shan State. *Mem. Geol. Sur. India*, 39, 2:1-379, 13 tab, Calcutta.

Myint Lwin Thein, 1973. The Lower Paleozoic Stratigraphy of the Western part of Southern Shan State, Burma; *Geol. Soc. Malaysia Bull. N.6*, p. 143-163.

Myint Thein, 1983. Geology of the Kywetnapa-Letpangon Area, Patheingyi and Maymyo Townships, M.Sc. Thesis (Unpublished), Department of Geology, University of Mandalay.

Pascoe, E., H., 1950, 1959, 1964. *Manual of the Geology of India and Burma*. 1st – 3rd ed. Y 83pp, Calcutta. (Govt. of India press.).

Thaw Tint, 1974. The Stratigraphy of Ngwetaung-Taunggyun-Kywenadauk Area, Pyin Oo Lwin Township (Abstract); Unpublished Research paper, Geol. Dept., University of Mandalay.

Wanless, H.R., 1973. Microstylolites, bedding and dolomitization (Abstract), Anahein, *AAPG Bull*, vol. 57., p.811.

Win Win Kyi., 1990. Geology and minerals resources of the Pathin-Kyadwinye area, Pyin Oo Lwin Township; unpublished M.Sc Thesis, Geol. Dept., University of Mandalay. 100pp.

Zaw Min Thein, 1995. Geology and Stratigraphy of the Thanpyegin Nyaungni Area, Pyin Oo Lwin Township, unpublished, M.Sc., Thesis, Geol. Dept., University of Mandalay, P.67.

Microfacies Analysis of the Thitsipin Limestone Formation, Dattaw Area, Kyaukse District, Mandalay Region

Zaw Min Thein[1], Sandar Tun[2], Su Wai Hlaing[3]

Abstract

Most of the Permian units are well exposed in Myanmar. The Permian Thitsipin Limestone Formation occupies the Dattaw Taung range, the Nwalegauk Taung and the plains situated on the western and eastern sides of Dattaw Taung. This unit is mainly composed of dark grey to bluish grey, commonly massive but locally poorly bedded, pure limestones with calcite veins and the great numbers of fusulinids, bryozoa, corals and brachiopods. Seven microfacies have been established in the Thitsipin Limestone Formation. The lithologic type and faunal content indicate that the Thitsipin Limestone Formation have been accumulated in a middle shelf environment, especially shoal, reef and reef mound.

Keywords: Thitsipin Limestone Formation, Microfacies, Middle shelf environment, Shoal, Reef and reef mound

Introduction

The Dattaw-Nattabin area is located at about 8 miles ENE of Kyaukse. It is situated between North Latitudes 21° 36' to 21° 43' and East Longitudes 96°15' 30" to 96° 17' 30". It is about 9 miles long in N-S direction and 2 miles wide in E-W direction, and thus, the total coverage of the area is approximately 18 square miles. One-inch topographic map of 93-C/6 was used as a base map. It is easily accessible by car and motor cycle throughout the year. The location map of the study area is shown in (Figure 1). The general geological mapping, stratigraphy of the Dattaw-Nattabin area and its environs were established by many authors (Pascoe, 1949; Maung Thein, 1971; Gramann et al., 1972; Maung Thein, 1973; Myint Lwin Thein, 1973; Garson et al., 1976; Hla Thein, 1976; Than Naing II, 1978; Myint Thein et al., 1984; Zaw Win,

1992; Mitchell, 1992; Than Soe Oo, 2001; Thura Oo et al., 2002; Zaw Min Thein, 2009).

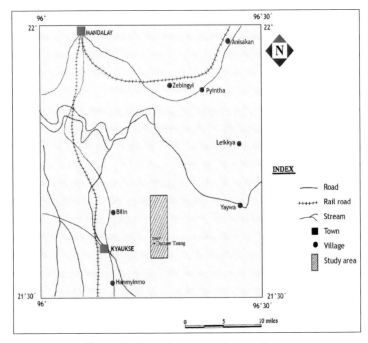

Figure (1) Location map of studied area

Purpose of Study

This research work describes the Permian Unit in terms of lithostratigraphy, facies analysis and develops a sedimentary facies model in Dattaw-Nattabin area. Microfacies of this stratigraphic unit have not been yet reported in this area. Therefore, the present studies make a close look at the occurrence of fauna and reconstruct the depositional environments of the study area by taking microfacies analysis.

Materials and Methods

In the laboratory, more than 130 specimens were cut and examined under the microscope. Each microfacies is classified on the basis of lithology, grain type, bedding, sedimentary structures and fossils. In the present area, seven microfacies

34

have been established in the Thitsipin Limestone Formation. On the basis of microfacies analysis, four distinct depositional settings can be recognized as shoal, reef and reef mound in middle shelf. The interpretation of depositional environment is supported by fluctuation in the diversity and abundance of the whole macrofossils throughout the measured succession.

Geological Setting

The Geological map of the Dattaw-Nattabin area which was modified based on the map prepared by Than Soe Oo, 2001 is shown in Figure (2) and the lithostratigraphic units of the Dattaw-Nattabin area have been classified in descending order.

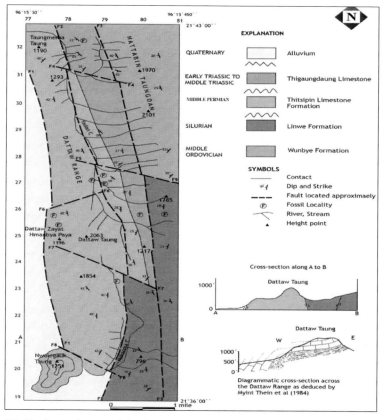

Figure (2) Geological map of the Dattaw-Nattabin area (Modified after Than Soe Oo, 2001)

35

Stratigraphic Column

    4. Thigaungdaung Limestone (Early to Middle Triassic)

          -----unconformity -----

    3. Thitsipin Limestone Formation (Middle Permian)

          -----unconformity -----

    2. Linwe Formation (Silurian)

    1. Wunbye Formation (Middle Ordovician)

Microfacies Analysis

In the Thitsipin Limestone Formation, the sections have been measured from Dattaw spring to Hmyanpya Pagoda (1196') and along Thapyeka track as shown in Figure (3) and (4). About 130 specimens were collected and examined under the microscope. On the basis of lithology, paleontology and sedimentary structures, it can be subdivided into shoal association and reef and reed mound association.

Symbols of Stratigraphic Columns

The followings are the symbols and legends that are used in the microfacies sections shown in Figure (3) and (4).

Legend

m : marls
M : mudstone
W : Wackestone
P : packstone
G : grainstone

calcareous siltstone
limestone
dolomatic limestone

algae
solitary coral
compound coral
forminifera
bivalve
gastropod
bryozoan
ostracod
echinoderm (crinoid)
brachiopod
ammonite

stromatolite
chert
lamination
calcisphere
fenestrae / birdseyes

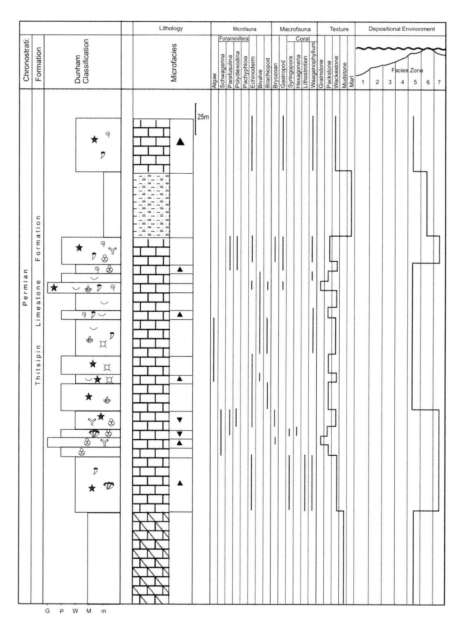

Figure (3) Generalized stratigraphic column of the Thitsipin Limestone Formation measured from the Dattaw spring to Hmyanpya Pagoda (1196 ft)

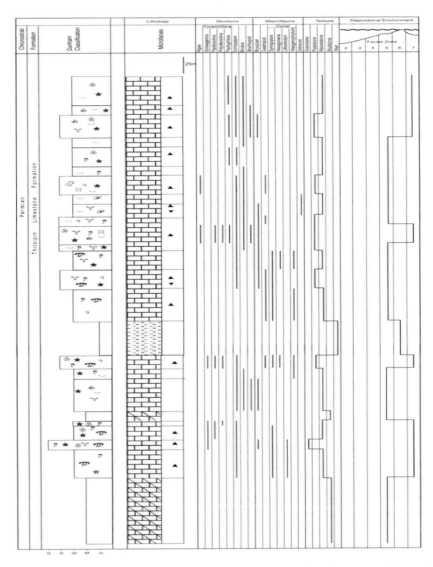

Figure (4) Generalized stratigraphic column of the Thitsipin Limestone Formation measured along the Thapyeka Track

Shoal Association

The shoal association that observed in studied area is shown in Figure (5) and (6).

Unfossiliferous calcisiltite

Lithology and grain type: Reddish brown to buff-coloured, fine-grained calcisiltite with scarce fossils, grains are usually subangular to subrounded and moderately sorted and are cemented by sparite and subordinate amount of iron.

Bedding and sedimentary structure: Well thin-bedded, but irregular in some places, gradationally changes to carbonates in both lower and upper boundaries, and characterized by coarsening upward nature.

Figure (5) Shoal Association: (A) The lower contact of the buff-coloured calcisiltite and micritic limestone in the middle part of Thitsipin Limestone Formation of Hmyanpya pagoda section (GPS: N21₀ 39′ 0.6″ and E96₀ 16′ 9.2″); (B) The upper contact of the buff-coloured calcisiltite and micritic limestone in the middle part of Thitsipin Limestone Formation of Thapyeka Track section (GPS: N21°39.639′ and E 96° 16.209′)

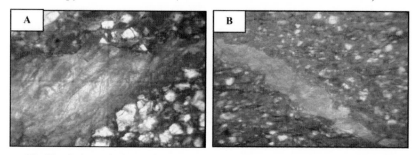

Figure (6) Shoal Association: The quartz-silt grains cemented with calcite spar and intercalation of calcite vein layer in quartz-silt grains (x40) (under PPL)

Reef and Reef Mound Association

The Reef and Reef Mound Association that observed in studied area is shown in Figure (7), (8), (9) and (10).

Massive, grey lime mudstone

Lithology and grain type: Bluish grey to dark grey, fine-grained, crystalline limestone. Fossils are rare. Fine-grained dolomite mosaics are also observed.

Bedding and sedimentary structure: Massive, often poorly-bedded but hard and compact, grey lime mudstone.

Thick- to very thick-bedded, bioclastic packstone-grainstone

Lithology and grain type: Light grey to bluish grey, finely crystalline and fossiliferous limestone. Fusulinids, crinoid stems, *Lithostrotion* sp., *Syringopora* sp., fenestrate bryozoa and echinoid plates are present.

Bedding and sedimentary structure: Thick-bedded to very thick-bedded bioclastic packstone-grainstone with solution pits and chert nodules.

Massive or poorly-bedded cherty bioclastic packstone-wackestone

Lithology and grain type: Bluish grey to dark grey, fine-to medium-grained cherty bioclastic packstone-wackestone. Usually rich in fusulinids but other bioclasts such as fenestrate bryozoa, *Syringopora* sp., *Waagenophyllum* sp., *Hexagoneria* sp. and crinoid stems are less common.

Bedding and sedimentary structure: Massive or poorly-bedded bioclastic packstone-wackestone. The chert stringers and nodules are more common.

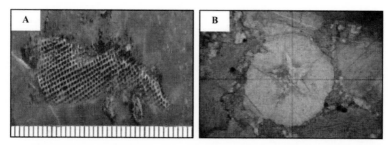

Figure (7) Reef and Reef Mound Association: (A) The *Fenestella* sp. in bioclastic packstone-wackestone facies of the Thitsipin Limestone Formation (GPS: N 21° 40.060' and E 96° 16.462); (B) Calcified crinoid stem and echinoderm in the crinoidal grainstone facies of the middle part of Thitsipin Limestone Formation (x40) (under PPL)

Medium-bedded, cherty bioclastic packstone

Lithology and grain type: Bluish grey to dark grey, fine-grained, cherty bioclastic packstone with fenestrate bryozoa, radiolarian and sponge specules.

Bedding and sedimentary structure: Medium-bedded, cherty bioclastic packstone. The chert nodules are common.

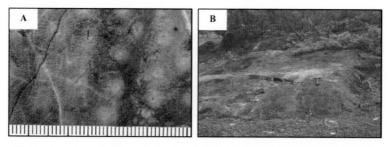

Figure (8) Reef and Reef Mound Association: (A) Massive or poorly bedded cherty bioclastic packstone-wackestone dominated by *Hexagoneria* sp. in the middle part of Thitsipin Limestone Formation. (GPS: N 21° 39' 11.3" and E 96° 16' 7.5"); (B) Massive lime mudstone facies of the lower part of Thitsipin Limestone Formation (GPS: N 21° 39' 12.1" and E 96° 15' 56.1")

Encrinite (Crinoidal grainstone)

Lithology and grain type: Light grey to bluish grey, crinoidal packstone composed of skeletal grains and lime mud matrix. Crinoid stems are relatively common and echinoid plates are present.

Bedding and sedimentary structure: Medium- to thick-bedded, encrinite (crinoidal grainstone).

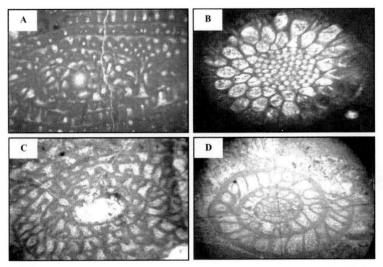

Figure (9) Reef and Reef Mound Association: Bioclastic packstone-grainstone facies in the lower part of Thitsipin Limestone Formation (A) *Polydiexodina* sp.; (B) *Streblascopora delicatula*; (C) *Parafusulina* sp.; (D) *Schwagerina* sp (x40) (under PPL)

Medium- to thick-bedded bioclastic wackestone

Lithology and grain type: Light grey to bluish grey, fine-grained, bioclastic wackestone. They are mainly composed of lime mud matrix, foraminifera and echinoid plates.

Bedding and sedimentary structure: Medium-to thick-bedded, bioclastic wackestone. The chert nodules are less common.

42

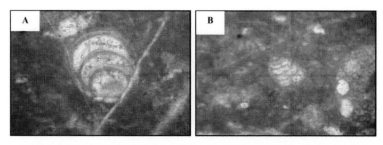

Figure (10) Reef and Reef Mound Association: (A) Bioclastic packstone -wackestone facies consisting of mainly the *Pachyphloia schwageri*; (B) *Pachyphloia aukurkoyi* in the upper part of Thitsipin Limestone Formation (x40) (under PPL)

Discussion

In the study area, the dolomitic limestone unit is firstly deposited in the lower part of the Thitsipin Limestone Formation. The dolomitic limestone unit is commonly bluish grey to dark grey in colour, massive crystalline limestone. The dolomitization of the lower part of the formation in this area may be fortuitous or may be the result of some lithological and sedimentological difference which rendered the older strata more vulnerable to the dolomitizing process. No fossil remain has been recognized from this dolomitic limestone unit. And then the micritic limestone unit bearing *Parafusulina-Polydiexodina* assemblage successively overlies the dolomitic limestone unit consists characteristically of dark grey, hard, compact massive limestone with black chert layers, patches and nodules. The micritic limestone is considered to have been deposited by chemical precipitation in a shallow, protected, quiet neritic environment. Fusulines are almost never found closely associated with saline deposits and in other types of clastic rocks (Garson, 1976). They thrived only in clear shallow seas at some distance from shore (Dunham, 1962, Tucker, 2001, Wilson, 1975). In the present area, the majority of shells are adult, free of abrasion or wear and generally similar in size. From these facts, it is inferred that they lived and accumulated on a quiet sea floor free from active agitation by waves and bottom currents. The above facts indicate that the micritic, fusuline limestone may be deposited under clear, quiet, shallow water, protected or sheltered neritic environment.

After the initial widespread carbonate deposition, reddish brown to buff-coloured calcisiltite is successively deposited. There is no such a feature of intense deformation at the contact. Unfortunately, no fossil evidence has been encountered. In microscopic study, quartz silt grains are cemented with calcite. Moreover, calcite veins layers alternate with quartz silt layers. This event is sediment aggradation and progradation during a still-stand and/or regressive period resulting in development of a shallowing upward sequence. The above facts indicate the calcisiltite may be deposited under shoal environment in agitated water. In the upper part, thick-bedded to massive, bluish to dark grey coloured, micritic limestone successively overlies the calcisiltite. This unit is closely associated with fenestellid bryozoans, corals, foraminifera, bivalves, brachiopods, echinoderms and ostracods. Brachiopods, echinoids, crinoids, nautiloids, and ammonites as well as many individual genera of other fossil groups indicate the normal marine salinity (Wilson & Jordan, 1983). On the basis of above faunal association, this unit may be deposited under shallow, warm and clear marine condition when the open shallow shelf (middle shelf) was under initial marine transgression (Figure 11).

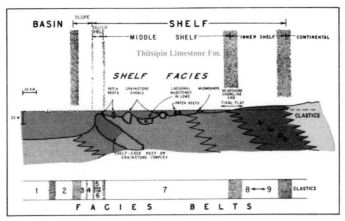

Figure (11) Profile of shelf facies showing the major subdivisions of shelf and faices (Modified from Wilson and Jordan, 1983)

Conclusion

The Dattaw-Nattabin area, covering part of the one-inch topographic map sheet 93 C/6, is located at about 8 miles ENE of Kyaukse, Mandalay Region. The Thitsipin Limestone Formation occupies the Dattaw Taung range, the Nwalegauk Taung and the plains situated on the western and eastern sides of Dattaw Taung. This unit is mainly composed of dark grey to bluish grey, commonly massive but locally poorly bedded, pure limestones with calcite veins and the great numbers of fusulinids, bryozoans, corals and brachiopods. Seven microfacies have been established in the Thitsipin Limestone Formation. On the basis of above faunal association, this unit may be deposited under shallow, warm and clear marine condition when the open shallow shelf (middle shelf) was under initial marine transgression.

Acknowledgements

I am indeed grateful to my supervisor Professor (Retired) U Bo San, Head of the Department of Geology, Kyaukse University and Pro-Rector Dr. Maung Maung for their enthusiastic guidance, various suggestions and close supervision.

References

Dunham, R.J. (1962). Classification of carbonate rocks according to depositional texture, in Ham, W.E., ed., Symposium classification of carbonate rocks: *Am. Assoc. Petroleum Geologists Memoir* **1**: 108-121.

Garson, M.S., Mitchell, A.H.G and Amos, B.J. (1976). The geology of the area around Neyaungga Yengan, Southern Shan State, Burma. Overseas. Mem. Inst. *Geol. Sci. London*; **2**, 70p.

Gramann, F., Fay Lain and Stoppel, D. (1972). Paleontological Evidence of Triassic Age for limestone from the Southern Shan and Kayah States of Burma: Hannover 1972.

Hla Thein (1976). The Stratigraphy of the Dattaw-Ye-Chaungzon Area, East of Kyaukse Township. MSc Thesis (unpublished), Geology Department, University of Mandalay.

Maung Thein (1971). Limestone Resources in the Kyaukse Area. Union of Burma. *Jour. Sci. Tech.*, **4**: 51-62.

Maung Thein (1973). A preliminary synthesis of the geological evolution of Burma with reference to tectonic development of Southeast Asia: *Geol. Soc. Malatsia Bull.* 6.

Mitchell, A.H.G., 1992. Late Permian-Mesozoic events and the Mergui Group Nappe in Myanmar and Thailand. *Journal of Southeast Asian Earth Sciences*, **7(23)**:165-178.

Myint Lwin Thein (1973). The Lower Paleozoic Stratigraphy of western part of the southern Shan State. *Geol. Soc. Malaysia, Bull.* **6,** 143 – 163.

Myint Thein, Bo San and Myint Thein (1984). Geological reinvestigation and the Jurassic-Cretaceous Sedimentation of the Area, East of Kyaukse-Belin, Mandalay Division; Unpublished report. University of Mandalay.

Pascoe, E.H. (1959). *A manual of the geology of India and Burma*: v.2, Govt. India Press, 3rd Ed, 1343 p.

Than Naing II (1978). Paleontology and geology of the Plateau Limestone Group, Dattaw Range, Kyaukse East. MSc Thesis (unpublished), Geology Department, University of Yangon.

Than Soe Oo (2001). Carbonate Facies and Foraminifera from the Plateau Limestone of Dattaw Taung, Kyaukse Township. MSc Thesis (unpublished), Geology Department, University of Yangon.

Thura Oo, Tin Hlaing and Nyunt Htay (2002). Permian of Myanmar. *Journal of Asian Earth Sciences*, 683 – 689.

Tucker, M.E., 2001. *Sedimentary Petrology.* Blackwell Scientific Publications, Oxford, 262p.

Wilson, J.L. (1975). *Carbonate Facies in Geologic History.* Spinger–Verlag, Berlin, Heidelberg, Germany.

Wilson, J.L., and Jorban, C. (1983). *Middle Shelf.* **In** Scholle, P.A., Bebout, D.G., Moore, C.H. (Eds), Carbonate Depositional Environments. *Mem. Am. Assoc. Pet. Geol.,* **33**: 298-343.

Zaw Min Thein (2009). Stratigraphy and sedimentology of the siliciclastic and carbonate rocks of the Dattaw-Nattabin area, Kyaukse District. PhD Dissertation (unpublished), Geology Department, University of Yangon. 87p.

Zaw Win (1992). *Lithostratigraphy and carbonate petrology of Lungyaw-Sakangyi area, Myittha-Ye-ngan Townships.* MSc Thesis (unpublished), Geology Department, University of Yangon. 174p.

Printed by Books on Demand GmbH, Norderstedt / Germany